On Wave Particle Duality and the Quantum of Action

Greg Feild

July 6, 2017

About the author:

I earned a Ph.D in experimental high energy physics from the Pennsylvania State University working on HERA at DESY in Hamburg, Germany studying photoproduction and deep inelastic scattering in electron-proton collisions.

I did my postdoctoral studies with Yale University working at Fermilab on the CDF experiment at the Tevatron. My primary research interest was particle hadronization in quarkonium production in proton-antiproton collisions.

"There's no use trying, one *can't* believe impossible things."

-- Alice to the White Queen

Abstract:

In this paper, we continue our interpretation
of the investigations of physics
as the science of mass in motion.

The universe is sinister,
but it is no longer spooky.

it's all about the oomph

The **NULL** hypothesis:

All physical interactions can be described by

Newton's Universal Laws + a Lagrangian

let's do physics!

Preface:

Whilst googling myself this afternoon (a bad habit I've developed recently),
I came across a little theory of gravity called Gravitoelectromagnetism (GEM).

I was like,

"This theory is brilliant!"

$$:)$$

"Genius!", I said.

The only thing the GEM equations (analogs of the Maxwell equations)
are missing is the mass current term, $\mathbf{J\_m}$, in the equation for electromagnetic
induction, which we introduced in "On Parity and Isospin".

Then I learned about Gravity Probe B.

Very exciting !

Particle in a box:

```
                                                                    ^
                                                                     \
                                                                      \
        |-----------------<-----------------------------\|
                                                              ^
                                                               \
                                                                \
        |-----------------<---------------------------\---|

        |-----------------<------------------------/---------|
                                                        /
                                                       /
                                                      /

node    |---------<----------------.---------------------------|

                                    ^
                                   /
                                  /
        |---------<-------/-------------------------------|

                              ^
                             /
                            /
        |-----<---/------------------------------------|

        |---------\--------------------------->------------|
                   \
                    \
                     \
```

artist's conception

Introduction:

In our last book, "A Critical Examination of Classical and Quantum Mechanical Waves", we proposed a mechanical model for the wavelike nature of the electron.

In our new model, the electron is an inertial (i.e. spin ½) blob of mass-energy/charge, spinning to the left. The axis of the electron spin and the principal axis of rotational inertia of the electron are aligned or 'projected' along the direction of motion of the electron and precess about this direction with a frequency proportional to the total mass-energy of the electron.

This precession frequency corresponds to the de Broglie wavelength of the electron;

$$v = E/h == m/h \qquad\qquad (1)$$

$$\lambda = h/p = h/mv \qquad\qquad (2)$$

with wavenumber

$$k = 2\pi/\lambda \qquad\qquad (3)$$

and so it takes 2π radians for electron spin to precess once about the direction of motion.

However, since the electron is a 'spinor', it takes 4π radians, or two revolutions of the spin vector, for the precessing *angular momentum vector* to return to its original value and helicity.

A free electron of fixed helicity, executes a 'polarization flip' every 2π radians, performing a 'complete revolution' every 4π radians.

This spin flipping is what gives a particle its oomph!

A free photon also performs a polarization flip every 2π radians.

That's the theory at least.

In addition, in our model, an electron's resistance to acceleration is a consequence of the conservation of angular momentum.

It's hard to shift about things that are spinning, and electrons are things that are spinning!

Particle in a box:

In "A Critical Examination of Classical and Quantum Mechanical Waves", we considered the classic example of the particle in a one dimensional box in light of our new theory.

A particle (e.g. an electron) in a box bounces back and forth between the walls. The particle follows a well defined path and physically traverses *all points* lying between the walls of the box.

Figure 1 shows the probability distribution for the state n = 2.

As we demonstrated in "Critical", the electron spin precesses around the direction of motion, performing a helicity flip every 2 pi radians, and completing one complete 'revolution' every 4 pi radians.

At the point x = L/2, the electron is undergoing a flip in helicity, where its spin is effectively zero, and hence it cannot interact with our experimental probe (e.g. a photon).

Similarly, at the walls of the box, the electron must perform a helicity flip as it bounces, exchanging a virtual photon with an electron in the wall of the box.

At the point x = L/4, the electron helicity is in 'full bloom' and it can interact with a real photon.

The electron oscillates harmonically between the ability to engage in real and 'virtual' interactions

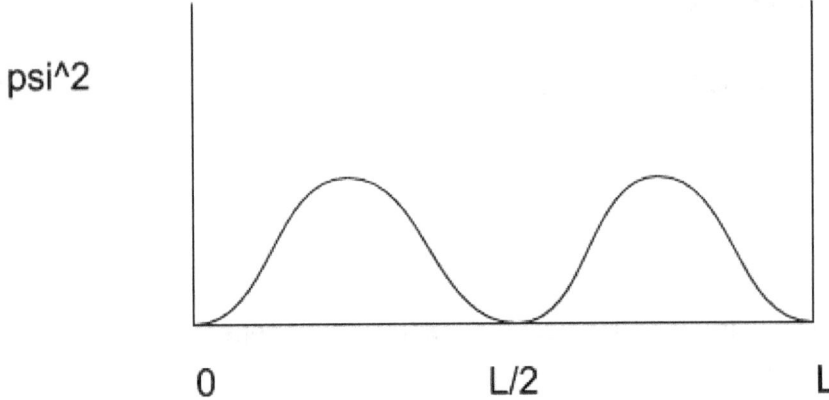

psi^2

0   L/2   L

Figure 1: Probability distribution for a particle in a box; n = 2.

Quantization:

Ultimately, quantization boils down to the satisfaction of boundary conditions.

Virtual photons can only attach to/interact with an electron when the electron has an effective helicity of zero (during its polarization flip) because virtual photons are spin zero.

This fixes the 'frequency' of the virtual photon coupling two interacting electrons, since virtual interactions, or the exchange of energy and momentum, can only occur when both electrons have helicity ~= 0.

When an electron reaches 'peak helicity', it is able to interact with (i.e. absorb and emit) real photons the most readily and the most efficiently.

Cosmic springs:

This view of quantization bolsters our conclusion that all interacting particles are constantly coupled by 'dedicated' virtual photons. These photons would vary in mass and frequency during an interaction in such a manner as to constantly satisfy the uncertainty principle.

Virtual photons *do not* radiate from a particle in a manner analogous to Faraday's lines of force. This radial representation is only valid for classical fields OR a source of real photons such as a radio antenna.

We argued in "Observations on the Quantum Mechanical Nature of Gravity", that particles do not send out 'feeler photons on spec', looking for something to couple to. This would require an infinite amount of time and energy, *and* chances are that most of these virtual photons would find *no target*.

When would a particle 'decide' it is time to give up and reel the virtual photon back in?

Entanglement:

The assumption behind things like entanglement and the EPR paradox is that particles do not have definite physical properties until they are measured!

Why anyone would think such a crazy thing is unclear.

In our theory, particles are real physical objects. They always have a definite position, momentum, and spin orientation.

We can only make probabilistic determinations of these quantities because particles are *very small*.

If one prepares two 'correlated' particles, for example, one with spin up and one with spin down, then one particle will always have spin up and the other spin down.

This will be true always and forever, no matter how far apart the particles may stray.

Quantum entanglement is not a thing.

The integral over all paths:

In the classical formulation of the Hamiltonian and Lagrangian approach to the dynamics of a system, one performs an integration over all possible paths in order to minimize the action.

One laughs and says, "This is crazy, I know the particle does no such thing!" However, the method gives the desired result, so one shrugs and moves on.

In quantum mechanics, of course, the integral over all paths formalism is accorded some mysterious and mystical significance. The particle is actually thought to follow all the possible paths available to it, and all at the same time !!!

We can be somewhat sympathetic to this view, since it was proposed in the bad old days of field theory.

Now it is another quantum boogeyman we must reject.

Hidden Variables:

We have discovered the hidden variables of quantum mechanics.

They are called spin, mass, and charge!

In our model, subatomic particle interactions are completely deterministic. In addition, there exist many physical and 'timing' constraints on these interactions, since the electron, as a spinor, can only interact at certain times, and in certain ways, as its helicity oscillates back and forth as it travels a *well defined* path.

In "Critical", we demonstrated that the Bohr radius of the electron for the hydrogen atom is *real, and,* that the electron travels a well defined, circular, and *closed* path as it orbits around the proton.

Reversibility:

In principle, the equations of classical and quantum mechanics are invariant under 'time reversal'; causing one to wonder why time seems to flow in only one direction in our physical universe.

In principle, the equations describing the earth's orbit around the sun are invariant under time reversal and space reflection.

In practice, one would need a really big 'tractor beam', and a whole lot of energy to decelerate the earth and send it off in the opposite direction; all the while trying to keep the earth from spiraling into the sun!

Time flows one way, into the future; unstoppable like a runaway freight train!

The single slit experiment:

In our model, the characteristic 'interference' pattern observed on the screen or photographic plate in a single slit diffraction experiment, is not due to two separate electron waves arriving at the screen out of phase and canceling each other out.

Instead, each separate electron arrives at the screen 'in or out of phase' for being able to interact with the photographic plate.

An individual electron can only darken the photographic plate if it arrives with the proper phase. Just like our particle in a box, the oscillating helicity of the electron has to be in the process of 'flipping polarization' when it arrives at the screen (i.e. the electron must have helicity ~zero to interact with the surface of the plate).

Electrons that do not successfully interact with the surface of the plate will carry on until they *do* find something to interact with.

We suggest a single slit experiment where the photographic plate is replaced with several layers of silicon, analogous to tracking detectors employed in high energy scattering experiments.

Our prediction is, of course, that the electrons which fail to contribute to the pretty pattern on the top layer, will show up in subsequent layers; perhaps displaying secondary pretty patterns.

The double slit experiment

We can employ similar arguments to explain double slit interference patterns, although in reality the situation is much more complicated.

In the single slit experiment, it does not really matter what happens before the slit. The slit is essentially the source of the electrons and the electrons have a characteristic spread determined by the width of the slit and the uncertainty principle.

For the double slit experiment, the environment of the electrons before the two slits *does* matter.

Although an electron can only pass through one slit or the other, it feels the 'potential' of both slits.

Remember, our 'free' electrons are actually exchanging virtual photons with *everything*. On their approach to the two slits, they are interacting with the wall housing the two slits, and they are interacting with the screen (which is their final destination) by virtual photon exchange via the two slits.

How this could actually be cast and analyzed in terms of potentials is unclear!

We suggest a similar double slit experiment, where the the two slits are replaced by two similarly sized and spaced metal plates, set to a positive retarding potential; thus, scattering the incoming electrons back toward the source and an appropriately placed screen.

The quantum of action:

There seems to be no entity called 'the quantum', and no fundamental physical quantity that corresponds to one 'quantum of action' (unless, of course, it turns out to be the electron neutrino!).

The formula relating the energy of a photon to its frequency, $E = h\upsilon$, shows that, in principle, the photon energy can take on *any* value.

We conclude the Planck constant (although a momentous discovery) is really nothing more than a glorified conversion factor describing the relationship between the frequency of a particle and the energy of a particle.

Any quantization of the energy of a photon, is ultimately due to the quantized nature of transitions in bound leptonic systems which result in photon emission.

The Lorentz force:

The complete, 'classical', relativistic, Lorentz force between two identical electrons is

$$\mathbf{F} = (G/c^4 - (e/m_e)^2(\mu/4\pi))(c/R)^2)(m_1 m_2 \mathbf{r} + (1/c^2)(\mathbf{p1xp2xr})) \tag{4}$$

where, of course, $\mathbf{F_1} = -\mathbf{F_2}$.

Newton's first law tells us

$$\mathbf{F_1} = m_1 \mathbf{a_1} \tag{5}$$

If we compare equations (4) and (5) we can see that the acceleration of a particle is *independent of its mass*.

$$\mathbf{a_1} = \text{Function}(m_2, \mathbf{R}) \tag{6}$$

This is a general result that we used to assume applied only to the gravitational interaction.

Now, *that* is mind bending physics!

The bottom line:

Particle interactions are all about balance.

Energy and momentum interact through the exchange of energy and momentum; conserving energy and momentum.

Facetious, but true!

Why this 'motion'? Why this constant shuffling?
Why this endless shuttling of energy and momentum
back and forth; to and fro?

Is there a method to this madness?

Is there some equilibrium to be reached?

Our current model indicates that 'the goal' is a balance between the mass-energy of a particle and the kinetic energy of a particle; or more properly and precisely, the balance between the kinetic and 'potential' energies of a system of particles.

$$\text{goal of universe} \; == \; m - m_0 \; = \; \text{kinetic energy} \; = \; \text{const.} \; == \; 42 \qquad (7)$$

                    of course!   :)

The answer is 42.

conclusion:

just be cool.

physics is fun!

:)

gilding the lily: : (

physics sells itself

Our troubled times:

When a physical theory, that make *no* physical sense,
is labeled 'mind bending' rather than discarded;

The mind boggles.

If a physical theory bends your mind,
it is wrong;
or at least, incomplete.

Correct theories straighten things out!

It is always better to say, "I don't know", than to spout some gibberish.

Similarly, one should always say "our current model says", rather than
"we now know" (e.g. space is curved, nucleons are made of quarks, etc., etc.);
even now, now that we have a Final Theory of Everything !

This is how reasonable, thoughtful, and *responsible* people use words.

Word.

References:

Modern Elementary Particle Physics
Gordon Kane

Quantum Physics
Rolf G. Winter

Gauge Theories in Particle Physics
I. J. R. Aitchison and A. J. G. Hey

Quarks and Leptons: An Introductory Course in Modern Particle Physics
Francis Halzen, Alan D. Martin

Symmetries and Group Theory in Particle Physics
Giovanni Costa, Gianluigi Fogli

and

Elementary Modern Physics
Richard T. Weidner, Robert L. Sells

a universal feild theory   :)

Books by Greg Feild:

the pentateuch

1. "A quantum mechanical theory of gravitational interactions"
   CreateSpace Independent Publishing, 8/29/2016

2. "Observations on the quantum mechanical nature of gravity"
   CreateSpace Independent Publishing, 10/8/2016

3. "On gravitation and electric charge"
   CreateSpace Independent Publishing, 11/1/2016

4. "On spin, mass, and charge"
   CreateSpace Independent Publishing, 11/29/2016

5. "On angular momentum, acceleration, and absolute motion"
   CreateSpace Independent Publishing, 1/4/2017

the exegeses

6. "The Sinister Universe"
   CreateSpace Independent Publishing, 3/1/2017

7. "On Parity and Isospin"
   CreateSpace Independent Publishing, 4/11/2017

8. "Reflections on the Sinister Universe"
   CreateSpace Independent Publishing, 5/12/2017

the hermeneutics

9. "On Current Physics"
   CreateSpace Independent Publishing, 6/11/2017

10. "A Critical Examination of Classical and Quantum Mechanical Waves"
    CreateSpace Independent Publishing, 6/18/2017

do physics!

Notes:

How I Did It: :)

Well . . . we certainly did not start from scratch!

All the pieces of the puzzle had been gathered;
they just needed to be put together.

The realization of the neutrino mass was the
tipping point; vaulting us all into the realm of

the *adjacent possible*.

peace out